Math Mammoth Grade 7 Tests and Cumulative Reviews

for the complete curriculum
(Light Blue Series)

Includes consumable student copies of:

- Chapter Tests
- End-of-year Test
- Cumulative Reviews

By Maria Miller

Copyright 2015 - 2020 Taina Maria Miller
ISBN 978-1-942715-26-9

EDITION 9/2020

All rights reserved. No part of this book may be reproduced or transmitted in any form or by any means, electronic or mechanical, or by any information storage and retrieval system, without permission in writing from the author.

Copying permission: For having purchased this book, the copyright owner grants to the teacher-purchaser a limited permission to reproduce this material for use with his or her students. In other words, the teacher-purchaser MAY make copies of the pages, or an electronic copy of the PDF file, and provide them at no cost to the students he or she is actually teaching, but not to students of other teachers. This permission also extends to the spouse of the purchaser, for the purpose of providing copies for the children in the same family. Sharing the file with anyone else, whether via the Internet or other media, is strictly prohibited.

No permission is granted for resale of the material.

The copyright holder also grants permission to the purchaser to make electronic copies of the material for back-up purposes.

If you have other needs, such as licensing for a school or tutoring center, please contact the author at
https://www.MathMammoth.com/contact.php

Contents

Test, Chapter 1 5
Test, Chapter 2 9
Test, Chapter 3 13
Test, Chapter 4 17
Test, Chapter 5 23
Test, Chapter 6 29
Test, Chapter 7 35
Test, Chapter 8 39
Test, Chapter 9 45
Test, Chapter 10 51
Test, Chapter 11 57
End-of-the-Year Test 63

Using the Cumulative Reviews 89
Cumulative Review Chapters 1-2 91
Cumulative Review Chapters 1-3 93
Cumulative Review Chapters 1-4 95
Cumulative Review Chapters 1-5 97
Cumulative Review Chapters 1-6 99
Cumulative Review Chapters 1-7 103
Cumulative Review Chapters 1-8 107
Cumulative Review Chapters 1-9 111
Cumulative Review Chapters 1-10 115
Cumulative Review Chapters 1-11 119

Grade 7, Chapter 1

End-of-Chapter Test

Instructions to the student:

Do **not** use a calculator. Answer each question in the space provided.

Instructions to the teacher:

You can give partial credit for partial solutions. The total is 21 points, so divide the student's score by 21 and multiply by 100 to get a percent score. For example, if the student scores 17, divide 17 by 21 to get 0.8095. The percent score is 81%.

Question	Max. points	Student score
1	2 points	
2	2 points	
3	1 point	
4	2 points	
5	2 points	

Question	Max. points	Student score
6	3 points	
7	4 points	
8	2 points	
9	3 points	
TOTAL	21 points	/ 21

Chapter 1 Test

1. Write an expression with three terms. The coefficient of the first term is 2 and of the second term is 5. The last term is the constant 9. The variable part of the first term is *s* squared, and the variable of the second term, *t*.

2. Evaluate the expressions.

a. $2(7-x)^2$, when $x = 2$	b. $\dfrac{1}{g} + \dfrac{g+1}{3}$, when $g = 6$

3. Name the property of arithmetic illustrated by the fact that $5(z + 4)$ is equal to $5z + 20$.

4. Draw a diagram of two rectangles to illustrate that the product $5(z + 4)$ is equal to $5z + 20$.

5. Write each expression as a product (factor it).

a. $7x + 21 = \underline{}(x + \underline{})$	b. $24k + 80 =$

6. Simplify the expressions.

a. $v + 5 + v + v + v$	b. $v \cdot 5 \cdot v \cdot v \cdot v$	c. $8x + 5 - 3x - 2$

7. Write an equation and solve it using guess and check.

   a. Seven times the quantity x minus one equals 14.

   b. Two less than x squared equals 23.

8. Write an expression for each situation.

   a. Abigail bought x bags of nuts for $3 a bag. She paid with a $50 bill. What was her change?

   b. A pair of jeans that costs p dollars is discounted by 1/10 of its price. What is the discounted price?

9. **a.** Write and simplify an expression for the total area of the shape.

   b. Evaluate your expression when $x = 2$ cm.

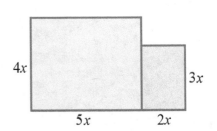

   c. Write an expression for the perimeter of the shape.

Grade 7, Chapter 2

End-of-Chapter Test

Instructions to the student:

Do **not** use a calculator. Answer each question in the space provided.

Instructions to the teacher:

You can give partial credit for partial solutions. The total is 36 points, so divide the student's score by 36 and multiply by 100 to get a percent score. For example, if the student scores 30, divide 30 by 36 to get 0.8333. The percent score is 83%.

Question	Max. points	Student score
1	2 points	
2	1 point	
3	4 points	
4	4 points	
5	2 points	
6	2 points	
7	4 points	

Question	Max. points	Student score
8	1 point	
9	3 points	
10	3 points	
11	4 points	
12	3 points	
13	3 points	
TOTAL	36 points	/ 36

Chapter 2 Test

1. Illustrate the sum −5 + 7 in two ways: (1) using counters, and (2) using a number line.

2. Write a sum where you add a number and its opposite.

3. Add or subtract.

| a. 11 + (−8) = | b. 11 − (−8) = | c. −11 − 8 = | d. −11 + 8 = |

4. Add or subtract.

| a. 61 + (−82) = | b. −55 − (−29) = | c. 43 − 189 = | d. 72 + (−99) = |

5. Explain a real-life situation for the calculation −80 m − 30 m = −110 m.

6. Write an addition or subtraction, and solve.

Aiden owed $21 on his credit card. Then he paid $15. Then he made a purchase for $35. Then he made another payment of $50. What is his balance now?

7. Simplify.

| a. $|-14| =$ | b. $-|-4| =$ | c. $-(-9) =$ | d. $-(13-8) =$ |

8. Allison's mom designed a point system for Allison where she would get positives for doing her chores and school work well and negatives for doing her chores and school work poorly

   On Tuesday, Allison "earned" five negative points. On Monday, her final tally had been 6 points. How many points better did Allison do on Monday than on Tuesday?

9. **a.** Write an expression for the distance between -2 and -18.

   b. Write an expression for the distance between x and 5.

   c. Evaluate the expression from (b) when $x = -2$.

10. Fill in the missing numbers.

| a. $-7 \cdot$ _____ $= -35$ | b. $2 \cdot$ _____ $= -100$ | c. $-50 \div$ _____ $= -10$ |

11. Divide and simplify to the lowest terms.

| a. $15 \div (-3) =$ | b. $-2 \div (-10) =$ |
| c. $-20 \div 6 =$ | d. $72 \div (-48) =$ |

12. Find the value of the expressions when $x = -3$ and $y = 4$.

| a. $xy - 2$ | b. $x^2 + 1$ | c. $-2(y - 5)$ |

13. Solve the equations by thinking logically.

| a. $-8y = 96$ $y =$ _____ | b. $-20a = -300$ $a =$ _____ | c. $36 \div w = -6$ $w =$ _____ |

Grade 7, Chapter 3

End-of-Chapter Test

Instructions to the student:

Use of a calculator with basic functions *is* allowed. Answer each question in the space provided.

Instructions to the teacher:

You can give partial credit for partial solutions. The total is 19 points, so divide the student's score by 19 and multiply by 100 to get a percent score. For example, if the student scores 14, divide 14 by 19 to get 0.7368. The percent score is 74%.

Question	Max. points	Student score
1	6 points	
2a	equation: 1 point	
	solution: 1 point	
2b	equation: 1 point	
	solution: 1 point	

Question	Max. points	Student score
3	3 points	
4	2 points	
5a	2 points	
5b	2 points	
TOTAL	19 points	/ 19

Chapter 3 Test

1. Solve. Check your solutions.

a.	$x + 8 = -13$	b.	$4 - (-2) = -y$
c.	$18 - x = -1$	d.	$2 - 6 = -z + 5$
e.	$\dfrac{x}{10} = -17 + 5$	f.	$-13 = \dfrac{c}{-7}$

Write an equation for each problem. Then solve it. Don't write just the answer.

2. **a.** Seven pounds of chicken costs $32.41. How much does one pound cost?

   b. Noah's suitcase is 4.6 kg heavier than Bill's. If Noah's suitcase weighs 28.7 kg, then how much does Bill's weigh?

3. Use the formula $d = vt$ to solve the problem.

A ferry travels at a constant speed of 18 km/h.
How long will it take to cross a river, a distance of 600 *meters*?

$d = v \quad t$
$\downarrow \quad \downarrow \quad \downarrow$

4. How far can you travel in 1 hour 25 minutes, bicycling at a constant speed of 15 km/h?

5. **a.** Find the average speed of an airplane that flies 2900 miles in 4 1/2 hours.

   b. Find the average speed of the same airplane in kilometers per hour.
   Use the conversion 1 mi = 1.609 km.

Grade 7, Chapter 4

End-of-Chapter Test

Instructions to the student:

Do **not** use a calculator. Answer each question in the space provided.

Instructions to the teacher:

You can give partial credit for partial solutions. The total is 43 points, so divide the student's score by 43 and multiply by 100 to get a percent score. For example, if the student scores 27, divide 27 by 43 to get 0.6279. The percent score is 63%.

Question	Max. points	Student score
1	5 points	
2	3 points	
3	2 points	
4	3 points	
5	2 points	
6	4 points	
7	3 points	
8	2 points	

Question	Max. points	Student score
9	2 points	
10	4 points	
11	4 points	
12	3 points	
13	2 points	
14	2 points	
15	2 points	
TOTAL	43 points	/ 43

Chapter 4 Test

1. Mark these numbers on the number line: $-\dfrac{2}{5}$, $-1\dfrac{4}{5}$, $-2\dfrac{1}{5}$, $-\dfrac{1}{10}$, $-1\dfrac{9}{10}$

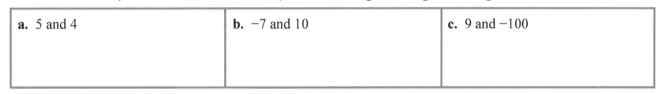

2. Form a fraction (numerator and denominator) from the two given integers. Then give it as a decimal.

a. 5 and 4	b. −7 and 10	c. 9 and −100

3. Write the decimals as mixed numbers.

a. 5.001	b. −2.0482

4. Write the fractions as decimals.

a. $-\dfrac{47}{10{,}000}$	b. $\dfrac{787}{10}$	c. $-\dfrac{5{,}624}{100}$

5. Which is more, $0.\overline{6}$ or 0.6?
   How much more?

6. Write as decimals, using a line over the repeating part (if any). Use long division.

a. $\dfrac{7}{6}$	b. $\dfrac{5}{36}$

7. Add or subtract.

a. $-1.26 - (-3.45)$	b. $1.8 - 3.25$	c. $-0.42 + 10.7 + (-9.8)$

8. Add or subtract.

a. $\dfrac{5}{9} + \left(-\dfrac{2}{3}\right)$	b. $-\dfrac{1}{10} - \dfrac{6}{9}$

9. Write the numbers in scientific notation.

   a. 25,600,000

   b. 7,810,000,000

10. Multiply or divide.

a. $-0.06 \cdot 0.05$	b. $(-0.5)^3$
c. $\dfrac{1}{3} \cdot \left(-5\dfrac{6}{11}\right)$	d. $-6\dfrac{1}{9} \div \left(-\dfrac{3}{4}\right)$

11. Divide *without* a calculator.

a. $1.5 \div 0.006$	b. $0.9 \div 0.011$

12. Give a real-life context for the calculation $\frac{1}{3} \cdot 12.75$. Then solve.

13. Find 15% of 3/4.

14. Two-thirds of a number is −5.66. What is the number?

15. Simplify these complex fractions.

a. $\dfrac{\frac{2}{5}}{4}$	b. $\dfrac{\frac{9}{10}}{\frac{1}{6}}$

Grade 7, Chapter 5

End-of-Chapter Test

Instructions to the student:

Do **not** use a calculator. Answer each question in the space provided.

Instructions to the teacher:

You can give partial credit for partial solutions. The total is 28 points, so divide the student's score by 28 and multiply by 100 to get a percent score. For example, if the student scores 19, divide 19 by 28 to get 0.6786. The percent score is 68%.

Question	Max. points	Student score
1	8 points	
2	equation: 1 point	
	solution: 2 points	
3	4 points	
4a	2 points	
4b	1 point	
4c	1 point	

Question	Max. points	Student score
5	2 points	
6	2 points	
7	1 point	
8a	2 points	
8b	1 point	
8c	1 point	
TOTAL	28 points	/ 28

Chapter 5 Test

1. Solve. Give your answers as fractions, mixed numbers, or whole numbers (not decimals).

a. $\quad -2 = 6x + 5$	**b.** $\quad 6x + 2x - 1 = -9x + 1$
c. $\quad \dfrac{3x}{5} = 24$	**d.** $\quad \dfrac{y}{3} - 21 = -5$

2. Ethan purchased 24 cookies and a loaf of bread for a total of $6.85. He didn't pay attention to the cost of the cookies but he remembered that the bread cost $3.25. Find the cost of one cookie by writing an equation and solving it.

3. Solve the inequalities and plot their solution sets on a number line. Write appropriate numbers for the tick marks yourself.

a. $3x + 5 < 68$

b. $10x - 17 \geq 103$

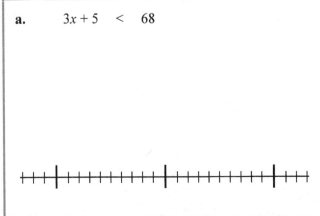

4. Abner got a building permit for a shed that limits its height to a maximum of 9.5 ft above the surface of the ground. He poured an above-ground foundation that was 4 inches thick, and the flat roof that he is going to add will extend 6 inches above the height of the wall. He is going to build with concrete blocks that are 8 inches tall (including the mortar).

   a. Write an inequality to calculate how many rows of block he can lay without the shed exceeding the maximum height permitted.

   b. Solve the inequality.

   c. Draw a number line and plot the solution set.

5. Plot the equations.

   a. $y = x + 4$

   b. $y = -3x + 5$

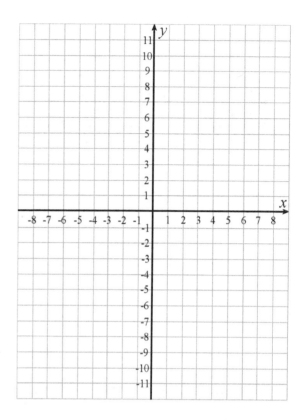

6. **a.** Draw any line that has a slope of -2.

   b. Draw a line that has a slope of 1/2 and that goes through the point (1, 3).

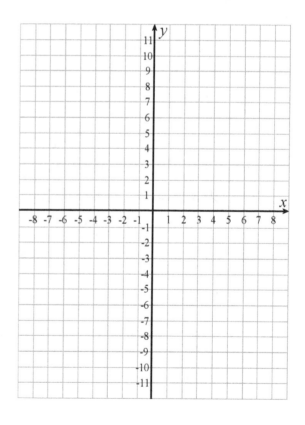

7. Determine the slope of this line from the table or from its graph.

x	0	1	2	3	4	5	6	7
y	20	30	40	50	60	70	80	90

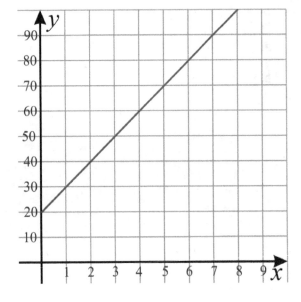

8. Leah runs at a constant speed of 4 m/s for 100 meters.

   a. Plot a graph for the distance Leah runs.

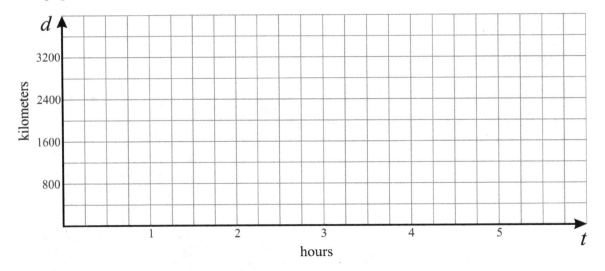

   b. How long does she take to run the 100 meters?

   c. Write an equation relating the distance (d) Leah has run and the time (t) that has passed.

Grade 7, Chapter 6

End-of-Chapter Test

Instructions to the student:

A basic calculator is allowed. Answer each question in the space provided.

Instructions to the teacher:

You can give partial credit for partial solutions. The total is 22 points, so divide the student's score by 22 and multiply by 100 to get a percent score. For example, if the student scores 20, divide 20 by 22 to get 0.90909. The percent score is 91%.

Question	Max. points	Student score
1	2	
2a	2	
2b	1	
3	3	
4	2	
5	2	
6	2	

Question	Max. points	Student score
7	2	
8a	1	
8b	1	
8c	1	
9a	1	
9b	2	
TOTAL		/ 22

Chapter 6 Test

You may use a calculator for all the problems in this test.

1. Chloe rode her bicycle 20 kilometers in 1 1/2 hours.
   Write a rate for her speed and simplify it to find the unit rate.

2. Mason poured 1/3 of an envelope of chocolate drink powder into 2/3 cups of water.

   a. Write the unit rate as a complex fraction and simplify it.

   b. What does the unit rate signify?

3. Write a proportion for the following problem and solve it.

   A bag of 52 kg of wheat costs $169.
   What would 21 kg of wheat cost?   ———— = ————

4. Solve the proportion by using cross-multiplication.

$$\frac{4.3}{S} = \frac{7.9}{12}$$

5. The figures are similar. Find the length of the side labeled with *x*.

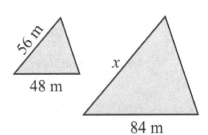

6. The aspect ratio of a television screen is 16:9 (width to height), and it is 63 cm high. What is its width?

7. A town map has a scale of 1:45,000.

   a. A street in this town is 850 m long. How long is that street on this map?

   b. How long in reality is a road that measures 5.4 cm on the map?

8. The graph on the right depicts the distance that a running fox covers as time passes.

   a. State the unit rate (including the units of measurement) for this situation.

   b. Plot the point that corresponds to the unit rate.

   c. Write an equation for the line.

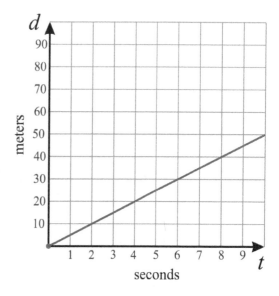

9. The equation S = 15t tells us the salary (S, in dollars) of a worker who works for *t* hours.

   a. What is the unit rate in this situation?

   b. Graph the equation S = 15t in the grid. Label the axes. Choose the scaling for the two axes so that the point corresponding for working for 10 hours will fit on the grid.

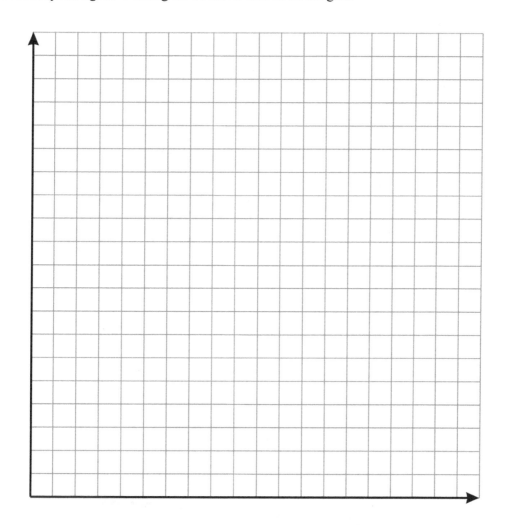

Grade 7, Chapter 7

End-of-Chapter Test

Instructions to the student:

A basic calculator is allowed. Answer each question in the space provided.

Instructions to the teacher:

You can give partial credit for partial solutions. The total is 16 points, so divide the student's score by 16 and multiply by 100 to get a percent score. For example, if the student scores 11, divide 11 by 16 to get 0.6875. The percent score is 69%.

Question	Max. points	Student score
1	3	
2	3	
3	1	
4	2	
5	2	

Question	Max. points	Student score
6	2	
7	1	
8	1	
9	1	
TOTAL		/ 16

Chapter 7 Test

You may use a basic calculator for all the problems in this test.
If not otherwise specified, give your answers that are percentages to the tenth of a percent.

1. The price of these items is changing. Find the new price or the discount percentage.

a. Price: $110 12% discount New price: $_____	b. Price: $5,000 2.4% increase New price: $_____	c. Price: $90 _____% discount New price: $59

2. The Jefferson family bought three tickets for children and two tickets for adults to the county fair. They got a 5% discount on the total purchase price before tax. Lastly, a 6.2% sales tax was added to the total. If the full price of a child's ticket is $10 and an adult's ticket is $20, find the total cost of tickets for the family.

3. This year the college has 1,210 students—an increase of 6.6% from last year. How many students did the college have last year?

4. Mary's dog weighed 25 kg, but then it got sick and lost 2.3 kg.

   a. What percentage of body weight did the dog lose?

   b. Mary weighs 58 kg. If Mary lost the same percentage of her body weight as what the dog did, how much would Mary weigh?

5. A rectangular playground area measures 5 m by 6.5 m. It is enlarged so that it becomes 7.2 m by 10 m. What is the percentage of increase in its area?

6. In 2010, the United States had 10,779,264 males and 10,320,257 females that were 0 to 4 years old. It also had 10,827,017 males and 11,282,003 females that were 50 to 54 years old.

   a. What is the percentage of more males than females in the age group 0-4 years?
   (Use relative difference.)

   b. What is the percentage of more females than males in the age group 50-54 years?
   (Use relative difference.)

7. A 12" pizza in Tony's Pizzeria costs $12.99 and in PizzaTown it costs $15.99.
   How many percent more expensive is the 12" pizza in PizzaTown than in Tony's Pizzeria?

8. Jacqueline deposited $2,500 into a savings account that pays a yearly interest rate of 4.4%.
   Calculate how much her account will contain after three years.

9. Michael borrowed $35,000 for ten years. At the end of those years he paid the bank back $65,800.
   What was the interest rate?

Grade 7, Chapter 8

End-of-Chapter Test

Instructions to the student:

Do **not** use a calculator for problems 1-6. You **may** use a basic calculator for problems 7-13. Answer each question in the space provided.

Instructions to the teacher:

You can give partial credit for partial solutions. The total is 30 points, so divide the student's score by 30 and multiply by 100 to get a percent score. For example, if the student scores 21, divide 21 by 30 to get 0.7. The percent score is 70%.

Question	Max. points	Student score
1	3	
2	2	
3	2	
4	2	
5a	1	
5b	3	
6	2	
7	3	

Question	Max. points	Student score
8a	2	
8b	1	
9	1	
10	1	
11	2	
12	2	
13	3	
TOTAL		/ 30

Chapter 8 Test

1. Find the measures of angles x, y, and z without measuring.

   $x =$ _____ $y =$ _____ $z =$ _____

2. Draw two angles that are complementary.

3. Write an equation for the unknown angle. Then solve it. Do not measure any angles.

   Equation for x: _____

   Solution:

4. Lines m and n are parallel. Find the measure of angle β without measuring.

5. An isosceles triangle has an 80° top angle and two 11-cm sides.

   a. Calculate the angle measure of the base angles.

   b. Draw the triangle.

6. Draw two lines that are perpendicular to each other using only a compass and a straightedge.

From this point on, you may use a basic calculator.

7. The "Yield" traffic sign is in the shape of an upside-down equilateral triangle. The image below shows the outline of its basic triangular shape, drawn at a scale of 1:10. Redraw it at a scale of 1:12.

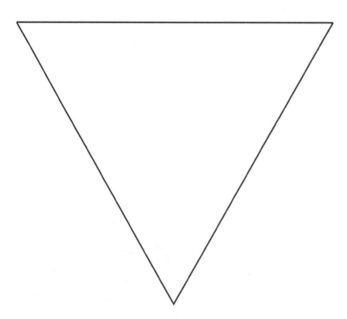

8. One acre is 43,560 square feet. Use that information and your knowledge of geometry to calculate the area of this trapezoidal plot:

   a. in square feet

   b. in acres, to the hundredth of an acre

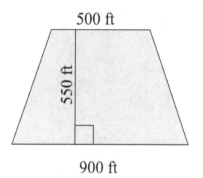

9. A rectangular prism is cut with a plane that is perpendicular to the prism's base. What figure is formed at the cross-section?

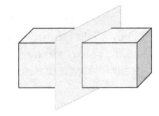

10. A rectangular pyramid is cut with a plane that is parallel to the pyramid's base. What figure is formed at the cross-section?

11. A large circular wall clock has a diameter of 40.0 cm. Find its area to the nearest ten square centimeters.

12. Calculate the volume of this water tank in <u>cubic meters</u>.

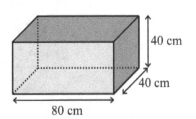

13. Calculate the surface area of this cylindrical bucket for toys to the nearest ten square centimeters.

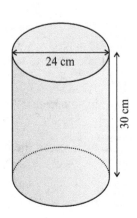

Grade 7, Chapter 9

End-of-Chapter Test

Instructions to the student:

Do **not** use a calculator for problems 1-3. You **may** use a basic calculator for problems 4-7. Answer each question in the space provided.

Instructions to the teacher:

You can give partial credit for partial solutions. The total is 25 points, so divide the student's score by 25 and multiply by 100 to get a percent score. For example, if the student scores 16, divide 16 by 25 to get 0.64. The percent score is 64%.

Question	Max. points	Student score
1	6	
2	2	
3	4	
4	2	

Question	Max. points	Student score
5	3	
6	3	
7	5	
TOTAL		/ 25

Chapter 9 Test

Do not use a calculator for problems 1-3 of the test.

1. Calculate the values of the square roots.

a. $\sqrt{1,000,000}$	b. $\sqrt{400}$	c. $\sqrt{1}$
d. $\sqrt{75-11}$	e. $\sqrt{10^4}$	f. $\sqrt{53^2}$

2. Do $\sqrt{-9}$ and $-\sqrt{9}$ have the same value? Explain.

3. Solve the equations. Round the answers to three decimals.

a. $s^2 - 17 = 19$	b. $5y^2 = 89 + 36$

From this point on, you may use a basic calculator.

4. Determine whether the lengths 13.4 m, 7 m, and 10.2 m form a right triangle using the Pythagorean Theorem.

5. Find the length of the unknown side.

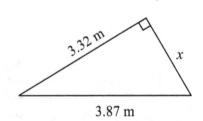

6. Find the length of a diagonal of a square with 50-cm sides.

7. Calculate the area of this shape to the nearest tenth of a square foot.

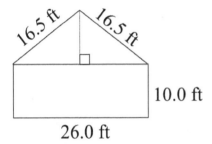

Grade 7, Chapter 10

End-of-Chapter Test

Instructions to the student:

A basic calculator is allowed. Answer each question in the space provided.

Instructions to the teacher:

You can give partial credit for partial solutions. The total is 22 points, so divide the student's score by 22 and multiply by 100 to get a percent score. For example, if the student scores 16, divide 16 by 22 to get 0.727272.... The percent score is 73%.

Question	Max. points	Student score
1	4	
2a	3	
2b	1	
2c	1	
2d	1	
3	3	

Question	Max. points	Student score
4a	1	
4b	3	
4c	1	
5	2	
6a	1	
6b	1	
TOTAL		/ 22

Chapter 10 Test

You may use a calculator for all the problems in this test.

1. You roll a number cube with numbers 1, 2, 3, 4, 5, and 6 printed on the faces. Find the probabilities as fractions.

    a. P(not 5)

    b. P(2 or 6)

    c. P(less than 9)

    d. P(not 2 nor 5)

2. Two spinners are spun.

    a. In the space below, draw a tree diagram showing all the possible outcomes of this experiment.

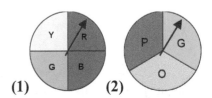
(1) (2)

   Then find the probabilities.

   b. P(yellow; purple)

   c. P(red or yellow; orange)

   d. P(not red; not orange)

3. Two dice are rolled. Find the probabilities of these events:

   a. You get a sum of six on the two dice.

   b. You get less than 3 on each dice.

   c. One dice is 6 and the other is not (in either order).

4. Logan and Alex tossed two coins 400 times.

   a. List all the possible outcomes when two coins are tossed just one time.

   b. Here are Logan's and Alex's results. Calculate and fill in the table the experimental and theoretical probabilities to the nearest tenth of a percent.

	Frequency	Experimental probability	Theoretical probability
TT	5		
TH	8		
HT	182		
HH	205		
TOTALS	400		

   c. Suggest a reason for the large discrepancy between the experimental and theoretical probabilities.

5. What is the probability of getting tails, tails, tails when you toss a coin three times in a row?

6. Lily and Grace placed some stuffed animals in a bag. Then they randomly pulled out one animal and put it back, and repeated this 120 times. Here are their results:

Animal	Frequency
Elephant	58
Giraffe	29
Bear	17
Cat	11
Bird	5
Totals	**120**

a. Based on their results, what is the approximate probability of pulling the cat out of the bag?

b. If this experiment was repeated 300 times, approximately how many times should they expect to get the bear?

Grade 7, Chapter 11

End-of-Chapter Test

Instructions to the student:

Do **not** use a calculator. Answer each question in the space provided.

Instructions to the teacher:

You can give partial credit for partial solutions. The total is 20 points, so divide the student's score by 20 and multiply by 100 to get a percent score. For example, if the student scores 15, divide 15 by 20 to get 0.75. The percent score is 75%.

Question	Max. points	Student score
1a	1	
1b	1	
1c	1	
1d	3	
2	6	

Question	Max. points	Student score
3	4	
4a	1	
4b	3	
TOTAL		/ 20

Chapter 11 Test

1. Researchers compared two different methods for losing weight by assigning 50 overweight people to use each method. The side-by-side boxplots show how many pounds people in each group lost.

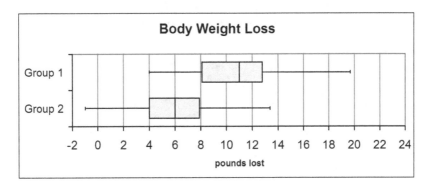

   a. Just looking at the two distributions, which group, if any, appears to have lost more weight?

   b. Which group, if any, appears to have a greater variability in the amount of weight lost?

   c. In group 2, there is one person whose weight loss was −1 pound. What does that mean?

   d. Is one of the weight loss methods significantly better than the other?

   If so, which one?

   Justify your reasoning.

2. Erica is studying the favorite hobbies of adults who live in a small town. She needs a sample of 120 people for her study. Below are listed four possible sampling methods that she could use. <u>Three</u> of the four methods would likely produce a biased sample. Explain which ones they are and why each method is biased.

Sampling method	Biased or not?
(1) Erica generates 120 random numbers between 1 and 1,000, and chooses the corresponding people that she meets on the main street of the town. If method (1) is biased, explain why:	
(2) Erica chooses randomly 120 people from a list of the town's residents. If method (2) is biased, explain why:	
(3) Erica places a box and her survey papers in the town's library, and anyone who wants to can fill in the survey questionnaire. If method (3) is biased, explain why:	
(4) Erica chooses randomly 120 people from a list of the town's taxpayers. If method (4) is biased, explain why:	

3. An ice cream shop surveyed its customers to find out which new flavors of ice cream to add to their selection. They obtained two samples using their customer database. Here are the results:

	Pineapple	Cappuccino	Peanut Butter	Kiwi	Blueberry	Totals
Sample 1	12	32	15	8	13	80
Sample 2	15	36	10	4	15	80

What can you infer from the results?

4. The two histograms show the age distributions of two groups of people. Below you find the means and the mean absolute deviations for both groups.

   a. How much do the means differ?

   b. Is the difference in the mean ages significant?

   Justify your reasoning.

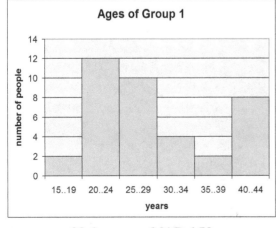

mean 28.6 years   MAD 6.78 years

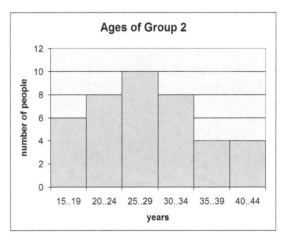

mean 27.55 years   MAD 6.205 years

Grade 7 (Pre-algebra) End-of-the-Year Test

This test is quite long, because it contains lots of questions on all of the major topics covered in the *Math Mammoth Grade 7 Complete Curriculum*. Its main purpose is to be a diagnostic test—to find out what the student knows and does not know about these topics.

You can use this test to evaluate a student's readiness for an Algebra 1 course. In that case, it is sufficient to administer the *first four sections* (Integers through Ratios, Proportions, and Percent), because the topics covered in those are prerequisites for algebra or directly related to algebra. The sections on geometry, statistics, and probability are not essential for a student to be able to continue to Algebra 1. The Pythagorean Theorem is covered in high school algebra and geometry courses, so that is why it is not essential to master, either.

Since the test is so long, I recommend that you break it into several parts and administer them on consecutive days, or perhaps on morning/evening/morning/evening. Use your judgment.

A calculator is *not* allowed for the first three sections of the test: Integers, Rational Numbers, and Algebra.
A basic calculator *is* allowed for the last five sections of the test: Ratios, Proportions, and Percent; Geometry, The Pythagorean Theorem, Probability, and Statistics.

The test is evaluating the student's ability in the following content areas:

- operations with integers
- multiplication and division of decimals and fractions, including with negative decimals and fractions
- converting fractions to decimals and vice versa
- simplifying expressions
- solving linear equations
- writing simple equations and inequalities for word problems
- graphs of linear equations
- slope of a line
- proportional relationships and unit rates
- basic percent problems, including percentage of change
- working with scale drawings
- drawing triangles
- the area and circumference of a circle
- basic angle relationships
- cross-sections formed when a plane cuts a solid
- solving problems involving area, surface area, and volume
- using the Pythagorean Theorem
- simple probability
- listing all possible outcomes for a compound event
- experimental probability, including designing a simulation
- biased vs. unbiased sampling methods
- making predictions based on samples
- comparing two populations and determining whether the difference in their medians is significant

If you are using this test to evaluate a student's readiness for Algebra 1, I recommend that the student score a minimum of 80% on the first four sections (Integers through Ratios, Proportions, and Percent). The subtotal for those is 118 points. A score of 94 points is 80%.

I also recommend that the teacher or parent review with the student any content areas in which the student may be weak. Students scoring between 70% and 80% in the first four sections may also continue to Algebra 1, depending on the types of errors (careless errors or not remembering something, versus a lack of understanding). Use your judgment.

You can use the last four sections to evaluate the student's mastery of topics in Math Mammoth Grade 7 Curriculum. However, mastery of those sections is not essential for a student's success in an Algebra 1 course.

My suggestion for points per item is as follows.

Question #	Max. points	Student score
Integers		
1	2 points	
2	2 points	
3	3 points	
4	6 points	
5	2 points	
6	3 points	
	subtotal	/ 18
Rational Numbers		
7	8 points	
8	3 points	
9	3 points	
10	2 points	
11	4 points	
	subtotal	/ 20
Algebra		
12	6 points	
13	3 points	
14	12 points	
15	2 points	
16a	1 point	
16b	2 points	
17	3 points	
18	4 points	

Question #	Max. points	Student score
19a	2 points	
19b	1 point	
20	8 points	
21	2 points	
22a	2 points	
22b	1 point	
	subtotal	/ 49
Ratios, Proportions, and Percent		
23	4 points	
24a	1 point	
24b	2 points	
24c	1 point	
24d	1 point	
25a	1 point	
25b	2 points	
26	2 points	
27	2 points	
28a	2 points	
28b	2 points	
29	2 points	
30	2 points	
31	2 points	
32	Proportion: 1 point Solution: 2 points	
33	2 points	
	subtotal	/ 31
SUBTOTAL FOR THE FIRST FOUR SECTIONS:		/118

Question #	Max. points	Student score
Geometry		
34a	2 points	
34b	2 points	
35	3 points	
36	2 points	
37	2 points	
38	2 points	
39a	1 points	
39b	3 points	
40a	2 points	
40b	2 points	
41	2 points	
42	3 points	
43a	2 points	
43b	2 points	
44a	2 points	
44b	2 points	
45a	2 points	
45b	1 point	
46a	1 point	
46b	2 points	
	subtotal	/ 40
The Pythagorean Theorem		
47	2 points	
48	2 points	
49	2 points	
50	3 points	
	subtotal	/9

Question #	Max. points	Student score
Probability		
51	3 points	
52a	2 points	
52b	1 point	
52c	1 point	
52d	1 point	
53	3 points	
54	3 points	
	subtotal	/14
Statistics		
55	2 points	
56a	1 point	
56b	2 points	
56c	2 points	
57	2 points	
58a	1 point	
58b	1 point	
58c	1 point	
58d	3 points	
	subtotal	/15
SUBTOTAL FOR THE LAST FOUR SECTIONS:		/78
TOTAL		/196

Math Mammoth End-of-the-Year Test - Grade 7

Integers

A calculator is not allowed for the problems in this section.

1. Give a real-life situation for the sum −15 + 10.

2. Give a real-life situation for the product 4 · (−2).

3. Represent the following operations on the number line.

   a. −1 − 4

   b. −2 + 7

   c. −2 + (−7)

4. Solve.

   a. −13 + (−45) + 60 = _____   **b.** −8 − (−7) = _____   **c.** 2 − (−17) + 6 = _____

   d. −3 · (−8) = _____   **e.** 48 ÷ (−4) = _____   **f.** (−2) · 3 · (−2) = _____

5. The expression | 20 − 31 | gives us the distance between the numbers 20 and 31.
   Write a similar expression for the distance between −5 and −15 and simplify it.

6. Divide. Give your answer as a fraction or mixed number in lowest terms.

   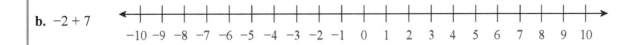

   a. 1 ÷ (−8)   **b.** −4 ÷ 16   **c.** −21 ÷ (−5)

Rational Numbers

A calculator is not allowed for the problems in this section.

7. Multiply and divide. For problems with fractions, give your answer as a mixed number in lowest terms.

a. $-\dfrac{2}{7} \cdot \left(-3\dfrac{5}{8}\right)$	**b.** $27.5 \div 0.6$
c. $-0.7 \cdot 1.1 \cdot (-0.001)$	**d.** $(-0.12)^2$
e. $\dfrac{\frac{3}{4}}{\frac{5}{12}}$	**f.** $\dfrac{5\frac{1}{2}}{-\frac{7}{8}}$
g. $-\dfrac{1}{6} \cdot 1.2$	**h.** $-\dfrac{2}{5} \div (-0.1)$

8. Write the decimals as fractions.

| a. 0.1748 | b. −0.00483 | c. 2.043928 |

9. Write the fractions as decimals.

| a. $-\dfrac{28}{10,000}$ | b. $\dfrac{2,493}{100}$ | c. $7\dfrac{1338}{100,000}$ |

10. Convert to decimals. If you find a repeating pattern, give the repeating part. If you don't, round your answer to five decimals.

| a. $\dfrac{7}{13}$ | b. $1\dfrac{9}{11}$ |

11. Give a real-life context for each multiplication or division. Then solve.

a. $1.2 \cdot 25$

b. $(3/5) \div 4$

Algebra

A calculator is not allowed for the problems in this section.

12. Simplify the expressions.

a. $7s + 2 + 8s - 12$	b. $x \cdot 5 \cdot x \cdot x \cdot x$	c. $3(a + b - 2)$
d. $0.02x + x$	e. $1/3(6w - 12)$	f. $-1.3a + 0.5 - 2.6a$

13. Factor the expressions (write them as multiplications).

a. $7x + 14$ =	b. $15 - 5y$ =	c. $21a + 24b - 9$ =

14. Solve the equations.

a. $\quad 2x - 7 = -6$	b. $\quad 2 - 9 = -z + 4$
c. $\quad 120 = \dfrac{c}{-10}$	d. $\quad 2(x + \tfrac{1}{2}) = -15$
e. $\quad \dfrac{2}{3}x = 266$	f. $\quad x + 1\dfrac{1}{2} = \dfrac{3}{8}$

15. Chris can run at a constant speed of 12 km/h. How long will it take him to run from his home to the park, a distance of 0.8 km?

   Remember to check that your answer is reasonable.

16. **a.** Which equation matches the situation?

   A pair of binoculars is discounted by 1/5 of their original price (*p*), and now they cost $48.

   | $\frac{w}{5} = 48$ | $\frac{4w}{5} = 48$ | $\frac{5w}{4} = 48$ | $w - 1/5 = 48$ | $w - 4/5 = 48$ | $5w - 4 = 48$ |

   b. Solve the equation to find the original price of the binoculars.

17. The perimeter of a rectangle is 254 cm. Its length is 55 cm. Represent the width of the rectangle with a variable and write an equation to solve for the width. Then solve your equation.

18. Solve the inequalities and plot their solution sets on a number line. Write appropriate multiples of ten under the bolded tick marks (for example, 30, 40, and 50).

a. $3x - 7 < 83$

b. $2x - 16.3 \geq 10.5$

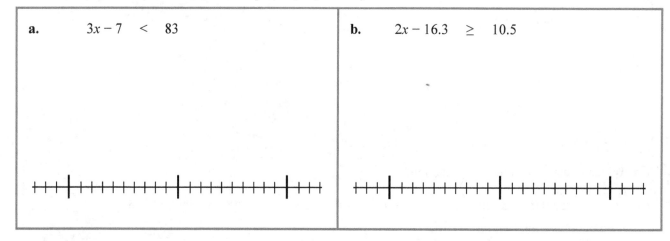

19. You need to buy canning jars. They cost $15 a box, and you only have $150 to spend. You also have a coupon that will give you a $25 discount on your total. How many boxes can you buy at most?

   a. Write an inequality for the problem and solve it.

   b. Describe the solution of the inequality in words.

20. *Solve.

a.	$9y - 2 + y = 5y + 10$	b.	$2(x + 7) = 3(x - 6)$
c.	$\dfrac{y + 6}{-2} = -10$	d.	$\dfrac{w}{2} - 3 = 0.8$

21. *Draw a line that has a slope of 1/2 and that goes through the point (0, 4).

22. a. *Draw the line $y = -2x + 1$.

    b. *What is its slope?

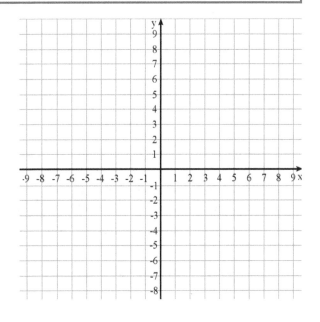

Ratios, Proportions, and Percent

You may use a basic calculator for all the problems in this section.

23. (1) Write a unit rate as a complex fraction. (2) Then simplify it. Be sure to include the units.

   a. Lily paid $6 for 3/8 lb of nuts.

   b. Ryan walked 2 ½ miles in 3/4 of an hour.

24. The graph below shows the distance covered by a moped advancing at a constant speed.

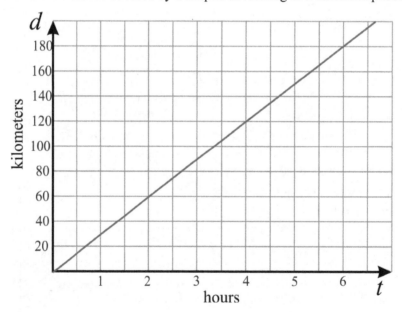

   a. What is the speed of the moped?

   b. Plot on the line the point that corresponds to the time $t = 4$ hours. What does that point signify in this context?

   c. Write an equation relating the quantities d and t.

   d. Plot the point that corresponds to the unit rate in this situation.

25. A Toyota Prius is able to drive 565 miles on 11.9 gallons of gasoline (highway driving). A Honda Accord can drive 619 miles on 17.2 gallons of gasoline (highway driving). (Source: Fueleconomy.gov)

   a. Which car gets better gas mileage?

   b. Calculate the difference in costs if you drive a distance of 300 miles with each car, if gasoline costs $3.19 per gallon.

26. Sally deposits $2,500 at 8% interest for 3 years.
    How much can she withdraw at the end of that period?

27. A ticket to a fair initially costs $10. The price is increased by 15%. Then, the price is decreased by 25% (from the already increased price). What is the final price of the ticket?

28. In December, Sarah's website had 72,000 visitors. In December of the previous year it had 51,500 visitors.

   a. Find the percentage of increase to the nearest tenth of a percent in the number of visitors her website had for that year.

   b. If the number of visitors continues to grow at the same rate, about how many visitors (to the nearest thousand) will her site have in December of the following year?

29. Alex measured the rainfall on his property to be 10.5 cm in June, which he calculated to be a 35% increase compared to the previous month. How much did it rain in May?

30. A square with sides of 15 cm is enlarged in a ratio of 3:4. What is the area of the resulting square?

31. How long is a distance of 8 km if measured on a map with a scale of 1:50,000?

32. Write a proportion for the following problem and solve it.

   600 ml of oil weighs 554 g.
   How much would 5 liters of oil weigh?   =

33. A farmer sells potatoes in sacks of various weights. The table shows the price per weight.

Weight	5 lb	10 lb	15 lb	20 lb	30 lb	50 lb
Price	$4	$7.50	$9	$12	$15	$25

   a. Are these two quantities in proportion?

   Explain how you can tell.

   b. If so, write an equation relating the two and state the constant of proportionality.

Geometry

You may use a basic calculator for all the problems in this section.

34. The rectangle you see below is Jayden's room, drawn here at a scale of 1:45.

    a. Calculate the area of Jayden's room in reality, in square meters.
    Hint: measure the dimensions of the rectangle in centimeters.

    b. Reproduce the drawing at a scale of 1:60.

    Scale 1:45

35. A room measures 4 ¼ in. by 3 ½ inches in a house plan with a scale of 1 in : 3 ft. Calculate the actual dimensions of the room.

36. Calculate the area of a circle with a diameter of 16 cm.

37. Calculate the circumference of a circle with a radius of 9 inches.

38. Draw a triangle with sides of 8 cm, 11 cm, and 14.5 cm using a compass and a ruler.

39. A triangle has angles that measure 36°, 90°, and 54°, and a side of 8 cm.

   a. Does the information given determine a unique triangle?

   b. If so, draw the triangle. If not, draw several different triangles that fit the description.

40. **a.** Write an equation for the measure of angle *x*, and solve it.

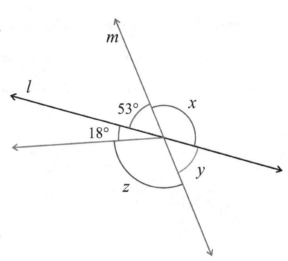

b. Write an equation for the measure of angle *z*, and solve it.

41. Calculate the measure of the unknown angle *x*.

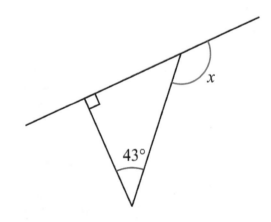

42. Describe the cross sections formed by the intersection of the plane and the solid.

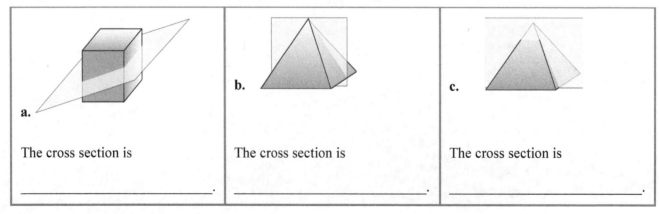

a. The cross section is _____.

b. The cross section is _____.

c. The cross section is _____.

43. **a.** Calculate the volume enclosed by the roof (the top part).

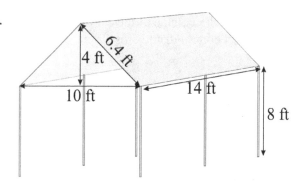

b. Calculate the total volume of the area covered by the canopy.

44. Two identical trapezoids are placed inside a 15 cm by 15 cm square.

a. Calculate their area.

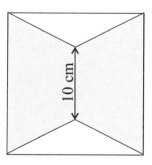

b. What percentage of the area of the square do the trapezoids cover?

45. **a.** *Find the volume of the cylindrical part of the juicer, if its bottom diameter is 12 cm and its height is 4.5 cm.

b. *Convert the volume to milliliters and to liters, considering that 1 ml = 1 cm³.

46. **a.** *How many cubic inches are in one cubic foot?

b. *The edges of a cube measure 3 ¼ ft. Calculate the volume of the cube in cubic inches.

The Pythagorean Theorem

You may use a basic calculator for all the problems in this section.

47. ***a.** What is the area of a square, if its side measures $\sqrt{5}$ m?

    *b. How long is the side of a square with an area of 45 cm²?

48. *****Determine whether the lengths 57 cm, 95 cm, and 76 cm form a right triangle. Show your work.

49. *****Solve for the unknown side of the triangle to the nearest tenth of a centimeter.

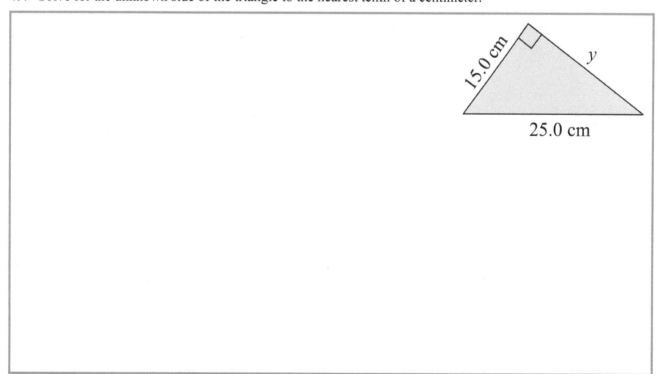

50. *You and your friends are at a river at point A. You suddenly remember you need something from home, which is at point C. So you decide to go home (distance AC) and then walk along the road (distance CB) to meet your friends, who will walk along the riverside from A to B.

If ABC is a right triangle, AC = 120 m, and CB = 110 m, how much longer is the distance (in meters) you will walk than your friends will walk?

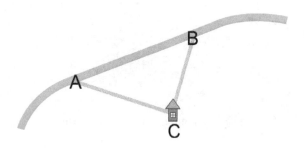

Probability

You may use a basic calculator for all the problems in this section.

51. You randomly pick one marble from these marbles.
    Find the probabilities:

    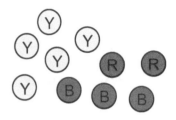

    a. P(not red)

    b. P(blue or red)

    c. P(green)

52. A cafeteria offers a main dish with chicken or beef. The customer then chooses a portion of rice, pasta, or potatoes, and a side dish of green salad, green beans, steamed cabbage, or coleslaw.

    a. Draw a tree diagram or make a list of all the possible meal combinations.

    A customer chooses the parts of the meal randomly. Find the probabilities:

    b. P(beef, rice, coleslaw)

    c. P(no coleslaw nor steamed cabbage)

    d. P(chicken, green salad)

53. John and Jim rolled a die 1,000 times. The bar graph shows their results. Based on the results, which of the following conclusions, if any, are valid?

    (a) This die is unfair.

    (b) On this die, you will always get more 1s than 6s.

    (c) Next time you roll, you will not get a 4.

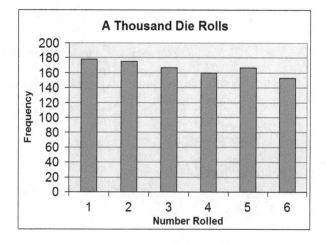

54. Let's assume that when a child is born, the probability that it is a boy is 1/2 and also 1/2 for a girl. One year, there were 10 births in a small community, and nine of them were girls. Explain how you could use coin tosses to simulate the situation, and to find the (approximate) probability that out of 10 births, exactly nine are girls. (You do not have to actually perform the simulation—just explain how it would be done.)

Statistics

You may use a basic calculator for all the problems in this section.

55. To determine how many students in her college use a particular Internet search engine, Cindy chose some students randomly from her class, and asked them whether they used that search engine.

    Is Cindy's sampling method biased or unbiased?

    Explain why.

56. Four people are running for mayor in a town of about 20,000 people. Three polls were conducted, each time asking 150 people who they would vote for. The table shows the results.

	Clark	Taylor	Thomas	Wright	Totals
Poll 1	58	19	61	12	150
Poll 2	68	17	56	9	150
Poll 3	65	22	53	10	150

    a. Based on the polls, predict the winner of the election.

    b. Assuming there will be 8,500 voters in the actual election, estimate to the nearest hundred votes how many votes Thomas will get.

    c. Gauge how much off your estimate might be.

57. Gabriel randomly surveyed some households in a small community to determine how many of them support building a new highway near the community. Here are the results:

If the community contains a total of 2,120 households, predict how many of them would support building the highway.

Opinion	Number
Support the highway	45
Do not support it	57
No opinion	18

58. Researchers compared two different methods for losing weight by assigning 50 overweight people to use each method. The side-by-side boxplots show how many pounds people in each Group lost.

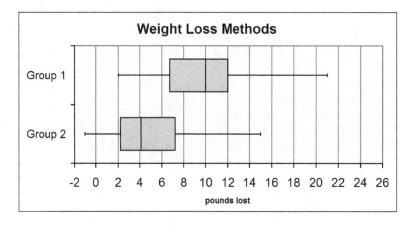

a. Just looking at the two distributions, which Group, if any, appears to have lost more weight?

b. Which Group, if any, appears to have a greater variability in the amount of weight lost?

c. In Group 2, there is one person whose weight loss was −1 pound. What does that mean?

d. Is one of the weight loss methods significantly better than the other?

If so, which one?

Justify your reasoning.

Using the Cumulative Reviews

The cumulative reviews practice topics in various chapters of the Math Mammoth complete curriculum, up to the chapter named in the review. For example, a cumulative review for chapters 1-6 may include problems matching chapters 1, 2, 3, 4, 5, and 6. The cumulative review lesson for chapters 1-6 can be used any time after the student has studied the curriculum through chapter 6.

These lessons provide additional practice and review. The teacher should decide when and if they are used. The student doesn't have to complete all the cumulative reviews. I recommend using at least three of these reviews during the school year. The teacher can also use the reviews as diagnostic tests to find out what topics the student has trouble with.

Math Mammoth complete curriculum also includes an easy worksheet maker, which is the perfect tool to make more problems for children who need more practice. The worksheet maker covers most topics in the curriculum, excluding word problems. Most people find it to be a very helpful addition to the curriculum.

The download version of the curriculum comes with the worksheet maker, and you can also access the worksheet maker online at

https://www.mathmammoth.com/private/Make_extra_worksheets_grade7.htm

Cumulative Review, Grade 7, Chapters 1-2

1. Rewrite each expression using a fraction line, then simplify.

a. $7 \div 8 \cdot 4$	b. $5 \cdot 2 \div 10 + 1$	c. $(10 + 3) \div (8 - 1)$

2. Evaluate the expressions. (Give your answer as a fraction or mixed number, not as a decimal.)

a. $\dfrac{x+2}{x-2}$, when $x = 21$	b. $3s^2 - 2t^2$, when $s = 10$ and $t = 3$

3. Name the property of arithmetic illustrated by the equation $(5x)y = 5(xy)$.

4. There are two broomsticks, one wooden and one metal.

   a. Choose two variables to denote the lengths of the two broomsticks.

   Let _____ be the length of the wooden broomstick.

   Let _____ be the length of the metal one.

   b. Write an equation that matches the sentence:
   "The wooden broomstick is 20 cm longer than the metal one."

5. a. Circle the equation that matches the situation.

   Let p be the normal price of one sun hat in a clothing store. The store owner decides to discount them by $5 each. A customer buys three sun hats, and the total cost is $16.80.

   $3(p - \$5) = \$16.80$ $p - \$5 = 3 \cdot \$16.80$

   $3p - \$5 = \$16.80$ $3(p - 0.5) = \$16.80$

   b. How much would one sun hat have cost before the discount? Solve this problem using any strategy. You don't have to use the equation.

6. Simplify the expressions.

a. $6p + 2 + 5p - 1$	b. $6p \cdot p \cdot 7$	c. $f \cdot 2f \cdot 2f \cdot f \cdot 3$

7. Simplify.

a. $|-71|$	b. $-|-2|$	c. $|-9+5|$	d. $-(-84)$
e. $|-9|+|-5|$		f. $|-9|-|5|$	

8. Write an inequality. Use negative integers where appropriate.

   a. This hill is at least 200 ft high.

   b. Liz owes more than $120.

   c. A maximum of 8 items per customer.

   d. The ride is only for children that are up to 120 cm tall.

9. Find the missing numbers. You can think of jumps on the number line.

a. $5 - \underline{} = {}^-2$	c. $2 + \underline{} = {}^-4$	e. ${}^-30 + \underline{} = {}^-40$	g. ${}^-51 + \underline{} = 0$
b. ${}^-1 - \underline{} = {}^-19$	d. $4 - \underline{} = 5$	f. $0 - \underline{} = {}^-49$	h. ${}^-9 + \underline{} = {}^-7$

10. Answer the questions about the pattern.

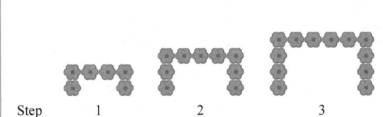

Step   1   2   3   4   5

   a. Draw steps 4 and 5.

   b. How do you see this pattern growing?

   c. How many flowers will there be in step 39?

   d. What about in step n?

Cumulative Review, Grade 7, Chapters 1-3

1. Find the root(s) of the equation $x^2 - x - 20 = 0$ in the set $\{-5, -4, -3, 3, 4, 5\}$.

2. Is division commutative? Explain your reasoning.

3. Solve.

a.	$\dfrac{x}{7} = -15$	b.	$11 = \dfrac{x}{-12}$	
c.	$7 - x = -3$	d.	$5 \cdot (-8) = -10x$	

4. Interpret the absolute value in each situation.

   a. The temperature is $-4°C$. $|4°C| = $ _____ $°C$

   Here, the absolute value shows _____

   b. The mountain is 2,500 ft tall. $|2,500 \text{ ft}| = $ _____ ft

   Here, the absolute value shows _____

5. Find the average speed in the given unit.

   a. Amanda swims 1 kilometer in 35 minutes.
      Give her average speed in kilometers per hour.

   b. You walk a distance of 1200 feet in 4 minutes.
      What is your average speed in miles per hour?

6. Add.

| a. $(-3) + (-6) + 5 + 1 =$ _____ | b. $14 + (-20) + (-31) + 11 =$ _____ |

7. Divide and simplify if possible.

| a. $12 \div (-5)$ | b. $-33 \div 15$ | c. $-2 \div (-9)$ |

8. a. Jerry's yard is a rectangle with one 500-ft side and a total area of 150,000 square feet. How long is the other side? Write an equation with an unknown and solve it.

   b. Sketch a square that is one yard long and wide. Use your sketch to figure out how many square feet are in one square yard.

   c. Lastly, convert the area of Jerry's yard into square yards.

Cumulative Review, Grade 7, Chapters 1-4

1. Are the expressions equal, no matter what values *n* and *m* have? If so, you don't need to do anything else. If not, provide a counterexample: specific values of *n* and *m* that show the expressions do NOT have the same value.

a. $(-n - 1) - m$ $-n - (1 - m)$	**b.** $\dfrac{x - y}{2}$ $\dfrac{x}{2} - \dfrac{y}{2}$

2. **a.** Sketch a rectangle with sides 3*s* and 8*s* long.

   b. What is its area?

   c. What is its perimeter?

3. Simplify the expressions.

a. $23r - 8r + 7r + 5$	**b.** $9p^2 + 8 - 3p^2$	**c.** $6y \cdot y \cdot 7y$

4. **a.** What is the total value, in cents, if Ashley has *n* quarters? Write an expression.

   b. Let's say the total value of her quarters is 875 cents. How many quarters does Ashley have? Write an equation and solve it.

5. Factor these sums (write them as products).

| a. $100x + 60 =$ | b. $24s - 4t - 8 =$ |

6. Consider the four expressions $67 + 28$, $(-67) + (-28)$, $(-67) + 28$, and $67 + (-28)$. Write these expressions in order from the one with **least** value to the one with **greatest** value.

7. **a.** Find the value of the expression $3 - x$ for at least six different values of x. Make a table to organize your work. Choose values of x in a pattern and notice the pattern in the values of $3 - x$.

   b. For which value of x will the expression $3 - x$ have a value of -2?

8. Solve the equation using the balance model. Write in the margin what operation you perform on both sides.

Balance	Equation	Operation to perform on both sides
	$4x - 1 = 7$	

Cumulative Review, Grade 7, Chapters 1-5

1. **a.** Which expression can be used to find the distance between x and 6?

| (i) $|x-(-6)|$ | (ii) $x-(-6)$ | (iii) $x-6$ | (iv) $|x-6|$ | (v) $|x+6|$ |

   b. Evaluate the expression when x is -23.

2. Justify the rule "A negative times a negative makes a positive" by filling in the missing parts of this proof based on the distributive property.

   (1) Substitute $a = -1$, $b = 1$, and $c = -1$ into the formula for the distributive property $a(b + c) = ab + ac$.

   ___ (___ + ___) = ___ · ___ + ___ · ___

   (2) The whole left side is zero because ___ + ___ = 0.

   (3) So the right side must equal zero as well.

   (4) On the right side, $-1 \cdot 1$ equals ___. So, $-1 \cdot (-1)$ must equal ___ so that the sum on the right side will equal zero.

   (5) Therefore, $-1 \cdot (-1)$ must equal ___.

3. Multiply.

| **a.** $(-7) \cdot 2 \cdot (-2)$ | **b.** $10 \cdot (-4) \cdot 7$ | **c.** $2 \cdot (-5) \cdot (-2) \cdot (-5)$ |

4. Below you see listed the minimum daily temperatures for one week. Calculate their average.

   $-8°C, -11°C, 2°C, 0°C, -3°C, -5°C, -1°C$

5. Multiply mentally.

a. $0.3 \cdot 2.5$	b. $-0.002 \cdot 0.008$	c. $-0.9 \cdot 50$
d. 0.8^2	e. $-4 \cdot 0.05 \cdot (-20)$	f. $(-0.3)^2$

97

6. Write the fractions as decimals.

a. $-\dfrac{61}{100{,}000}$	b. $\dfrac{9{,}807{,}200}{1000}$	c. $\dfrac{55{,}191}{1{,}000{,}000}$

7. Alex commutes 12 km to work every day. One day, his average speed going to work was 60 km/h and coming back 50 km/h. How long did it take Alex to commute that day?

8. Write in decimal form. Use long division, and calculate each answer to at least six decimal places. If you find a repeating pattern, give the repeating part. If you don't, round your answer to five decimals.

a. $2\dfrac{5}{24}$	b. $2.05 \div 7$	c. $5.6 \div 0.02$

9. Add the fractions.

a. $\dfrac{3}{5} + \left(-\dfrac{2}{3}\right)$	b. $-\dfrac{1}{2} + \left(-\dfrac{6}{9}\right)$

Cumulative Review, Grade 7, Chapters 1-6

1. Simplify the expressions.

a. $-3z - 9 + 7z + 2t$	b. $6x \cdot x \cdot (-7x)$	c. $6s \cdot s \cdot 4t$

2. Solve *without* a calculator.

a. 30% of $400	b. $\frac{2}{3} \cdot 6.9$ km	c. $0.08 \cdot \frac{1}{10}$

3. Solve. Check your solutions.

a. $2x - 6 = 9x - 8$	b. $3(x - 6) = -9x$
c. $8x = -\frac{3}{4}$	d. $1\frac{1}{6} + v = \frac{2}{9}$

4. Find the slope of each line.

   a.

   b.

   c.

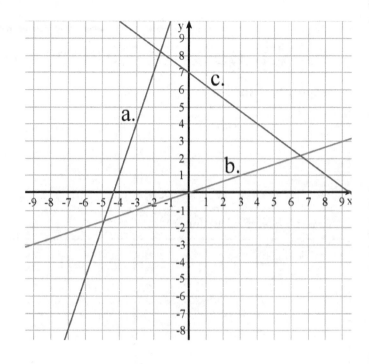

5. Write the numbers in scientific notation.

   a. 490,000,000

   b. 6,238,000,000

6. Write the numbers in numerical form.

   a. $2.08 \cdot 10^8$

   b. $1.293 \cdot 10^6$

7. a. Draw a line that has a slope of 2 and that goes through the point (0, 6).

   b. Draw a line that has a slope of −1/2 and that goes through the point (−4, 7).

   c. Draw a line that has a slope of 4/3 and that goes through the point (0, 1).

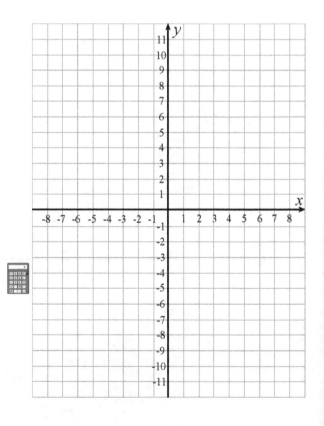

8. Cynthia took 14 minutes to bicycle from her home to a dentist appointment (a distance of 2.8 km) and 10 minutes to bicycle from there back home. Calculate her average speed for the entire trip.

9. A contractor has quoted you a price of $4.50 per square foot for a driveway. The driveway will be 40 ft long but you need to decide the exact width. You can afford to spend at most $1,600.

   a. Find the maximum width for the driveway.
   Hint: First find the maximum area for the driveway based on what you can afford.

   b. Write an equation for finding the maximum width of the driveway, and solve it.
   I realize you already know the answer from solving it in (a), but the purpose of this exercise is to let you practice how to write equations for real-life situations.

10. An airplane travels at a constant speed of 500 mi/h.

    a. Write an equation relating the distance (d) it has traveled and the time (t) that has passed.

    b. Plot your equation. Notice that you need to scale the d-axis.

    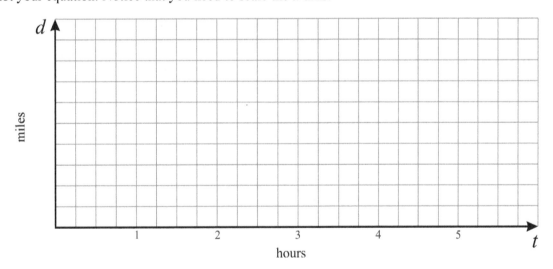

    c. How long will it take the airplane to travel 3,600 miles?

Cumulative Review, Grade 7, Chapters 1-7

1. Solve the inequalities and plot their solution sets on a number line. Write appropriate multiples of ten under the bolded tick marks (for example, 30, 40, and 50).

   a. $\quad 3y + 7 < 56$

   b. $\quad -5 + 6z \geq 175$

2. As a salesperson selling fine art paintings, you are paid a base salary of $180 per week plus $45 per painting. How many paintings do you need to sell in a week in order to earn at least $500?
   Write an inequality for the number of sales you need to make, solve it, and describe the solutions.

3. Matt got a really unreasonable answer for the problem below. Find what went wrong with his solution and correct it.

   Jim can swim 30 laps in a pool in 26 minutes.
   How many laps could he swim in 45 minutes?

   Matt's Answer: He could swim 30 laps.

   Solution:
   $$\frac{30 \text{ laps}}{26 \text{ min}} = \frac{L}{45 \text{ min}}$$
   $$26L = 30 \cdot 26$$
   $$26L = 780$$
   $$\frac{26L}{26} = \frac{780}{26}$$
   $$L = 30$$

4. **a.** Plot the equation $y = 2x - 1/2$.

   b. Plot the equation $y = -(1/3)x$.

   c. Plot the equation $y = 10 - 2x$.

5. Determine whether the point $(-2, 2)$ is on the line $y = -(1/2)x + 1$. Show your work.

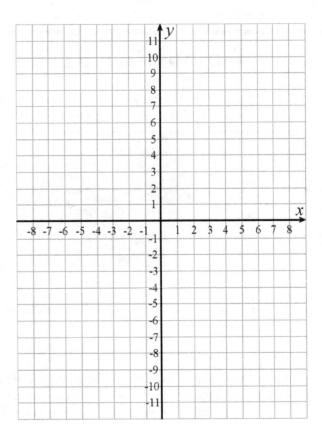

6. The two triangles are similar. How long is the unknown side?

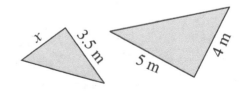

7. **a.** Draw a copy of this rectangle using the scale factor 2.5.

   b. What is the scale *ratio*?

8. **a.** Find the value of the expressions 2p and p − 4 for different values of p.

Value of p	2p	p − 4
−6		
−5		
−4		
−3		

Value of p	2p	p − 4
−2		
−1		
0		
1		

Value of p	2p	p − 4
2		
3		
4		
5		

b. Consider the patterns in the values of these expressions.
Is there any value of p for which 2p would be equal to p − 4?

c. For which values of p in the table is 2p more than p − 4?

d. For which value of p does the expression p − 4 have the value −2?

9. Solve (without a calculator).

a. $10 - \frac{5}{6} \cdot 2.7$

b. $0.4 \div \left(\frac{2}{9} + \frac{1}{3}\right)$.

10. Solve.

a. $13 + (-37) + (-8) =$

b. $-94 - (-8) =$

c. $20 - 60 + 90 =$

11. Find the missing numbers.

a. $6 \cdot \_\_\_\_ = -42$

b. $-72 \div \_\_\_\_ = 8$

c. $\_\_\_\_ \div (-12) = -4$

12. Find the value of the expressions when x = −5 and y = 2.

a. $1 - x^2$

b. $10xy$

c. $-3(x + y)$

Cumulative Review, Grade 7, Chapters 1-8

1. **a.** Sketch a rectangle with sides $5x$ and $6x$ long.

   b. What is its area?

   c. What is its perimeter?

2. A photo editing software was discounted by 2/5. The discounted price is $29.97.

   a. Find the original price using logical reasoning and/or a bar model.

   b. Choose a variable to represent the original price. Write an equation for the situation and solve it. Compare the solution steps of the equation to the way you solved the problem in (a).

3. Find the percentage of increase or decrease.

a. A flashlight that costs $9 is discounted so that now it costs $8.10. What percentage is it discounted?	**b.** A chair used to cost $20, but now it costs $26. What is the percentage of increase?

4. Mason got 16 points out of 21 in a quiz. What is his quiz score score, to the nearest tenth of a percent?

5. Mark pays 22.5% of his income in taxes. If he earns $2,350 in a month, find how much he has left after taxes.

6. A farmer gets paid $3 for a bushel of corn. A bushel is about 35.2 liters.

   a. How much does the farmer get for one liter of corn?

   b. Let P be the the amount of money the farmer gets and V be the amount of corn in liters. Write an equation relating the two variables.

   c. Plot your equation. Choose appropriate scaling for the P-axis.
   Hint: calculate how much the farmer gets for 100 liters and for 600 liters of corn.

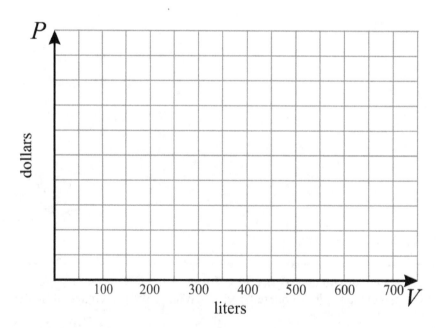

   d. Plot the point corresponding to 200 liters of corn.

   e. Plot the point corresponding to the farmer earning $50.

7. Study the problem below and Jane's solution for it. She figured that exactly half of the books were to be sent to book stores.

A printing press printed 1,500 copies of a book. 5/6 of those were printed as paperbacks and the rest were printed with hard covers. Now, 3/5 of the paperbacks need to be sent to various book stores. How many books is that?	Jane's calculation to solve this: $\dfrac{\cancel{3}^1}{\cancel{5}_1} \cdot \dfrac{\cancel{5}^1}{\cancel{6}_2} \cdot 1{,}500 = \dfrac{1}{2} \cdot 1{,}500 = 750$

   a. Is her answer correct?

   b. Is the way she calculated it correct? If not, correct it.

8. a. Draw a rectangle with an aspect ratio of 4:3 (width to height) so that the width is 6 cm.

   b. Then enlarge your rectangle using the scale ratio 3:5. Draw the enlarged rectangle.

   c. What is the aspect ratio of the enlarged rectangle?

Cumulative Review, Grade 7, Chapters 1-9

1. **a.** In the following problem, represent Jayden's salary with a variable and write an equation for the situation.

    Jayden pays 1/7 of his salary in taxes. If he paid $415 in taxes, how much was his salary?

    b. Solve the equation.

2. Factor these expressions (write them as products).

a. $7x + 21 = $ ___ (___ + ___)	**b.** $24w - 16 = $ ___ (___ - ___)
c. $-21t - 7 = -7($ ___ + ___)	**d.** $50a - 70b - 120 = $
e. $-55a + 30 = $	**f.** $-56y - 84 - 7x = $

3. Ava used 3 1/2 cans of paint to paint 2/3 of a room.

    a. Write the unit rate as a complex fraction and simplify it.

    b. Using paint at the same rate, how much more paint does Ava need to finish painting the room?

4. Evaluate the expressions.

a. $100 - x^2$, when $x = -2$	**b.** $\dfrac{2w}{w + 3}$, when $w = 1/2$

5. Solve.

a. $x - \dfrac{5}{6} = 7\dfrac{1}{3}$	b. $2\dfrac{1}{4} - w = 1\dfrac{2}{7}$
c. $5y = -\dfrac{4}{9}$	d. $v + \dfrac{1}{5} = -\dfrac{1}{12}$

6. The price of a standing lamp increased from $19 to $22.50. What was the percentage of increase?

7. Mason took a $1,500 loan at 9.8% annual interest rate to purchase a computer. He paid it back 1 1/2 years later. Calculate the total amount Mason paid at that point.

8. The table below shows how long it takes for a car to travel a distance of 120 km at different speeds.

Speed (km/h)	120	100	80	60	40	20
Time (h)	1	1.2	1.5	2	3	6

   a. Are these two quantities, speed and time, in proportion?

   Explain how you can tell that.

   b. If so, write an equation relating the two and state the constant of proportionality.

9. The four children in the Adams family have earned the following points in a computer game.

Child	Points
Chris	365
Grace	458
Hailey	602
Tony	553

   a. Comparing Grace and Hailey, how much better percentage-wise is Hailey doing than Grace? Use relative difference.

   b. Comparing Tony and Chris, how much better percentage-wise is Tony doing than Chris? Use relative difference.

10. **a.** Draw an equilateral triangle using only a compass and a straightedge. Make the side whatever length you wish. If you draw a small one, you can draw it here. Or, you can use blank paper.

    b. Draw the altitude into it.

    c. Use a centimeter ruler to measure what you need, and find its area to the nearest square centimeter.

11. A triangle has 45° and 60° angles and a 5.6-cm side between those angles. Does the information given define a unique triangle? If it does, write yes, and draw the triangle.

   If not, prove that it doesn't by sketching at least two non-congruent (different-shaped) triangles that satisfy the given conditions.

Cumulative Review, Grade 7, Chapters 1-10

1. Find the square roots (without a calculator).

| a. $\sqrt{1}$ | b. $\sqrt{64}$ | c. $\sqrt{10{,}000}$ | d. $\sqrt{400}$ |

2. The floors of a skyscraper are 11 feet apart. The bottom floor, which is actually the basement, is located at 9 feet below the ground.

   a. Write an expression that tells you the elevation of the nth floor.

   b. At which elevation is the 47th floor?

   c. Write an equation to find which floor is at an elevation of 288 ft, and solve it.

3. Explain how we can use the pictures on the right to show that the area of a circle is $A = \frac{1}{2}C \cdot r$, where C is the circumference and r is the radius of the circle.

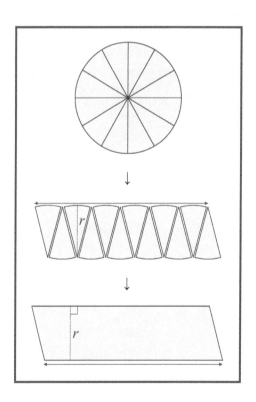

115

4. In an educational game, you earn 3 play coins for each 10 problems you complete.

   a. Give the unit rate as problems per coin.

   b. How many problems would you need to solve in order to earn 75 coins so you can purchase a phone for your virtual character in the game?

   c. In the same game, if you have solved 376 problems, how many coins have you earned?

5. Which covers more of the wall, a square clock with 30-cm sides or a circular clock with a diameter of 36-cm?   How much more?

6. Alison pays $14.95 per month for an internet service and she can use 200 gigabytes of bandwidth in that time. For any bandwidth she uses in excess of 200 gigabytes, she will pay $0.35 per gigabyte. She used 75% of her monthly bandwidth quota the first 18 days in September. If she continues to use bandwidth at the same rate, how much extra will she pay at the end of the month?

7. The two figures are similar. Calculate the length of the side marked with x.

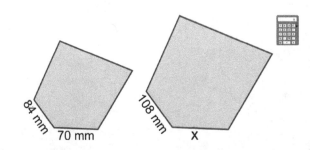

8. Solve for the unknown side of the triangle to the nearest tenth of a foot.

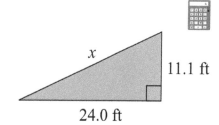

9. A hexagonal prism is 12 cm high. Its base is a regular hexagon with a side of 2.5 cm and an area of 16.2 cm².

   a. Calculate the surface area of the prism.
   Hint: Sketch the prism to help you.

   b. Calculate the volume of the prism.

10. You choose the digits for a two-digit number randomly from the digits 2, 3, 5, and 7 (prime numbers). Each digit can be used twice; for example, it is possible to make 55.

    a. What is the probability of making a number between 31 and 40?

    b. What is the probability of making a number where both digits are the same?

Cumulative Review, Grade 7, Chapters 1-11

You may use a basic calculator for all the problems in this lesson.

1. The line graph shows the number of candles that the candle factory sold in the years from 2010 to 2015. Note that the scale is given in "thousand candles."

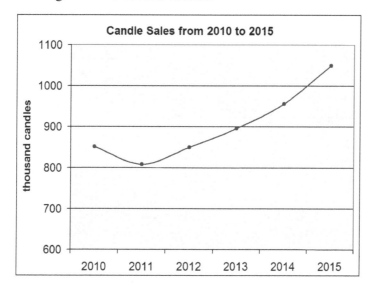

Estimating the amounts from the graph, calculate the approximate percentage increase in the candle sales from 2010 to 2015.

2. The shape on the left was scaled to become the shape on the right. Find the scale ratio, and then write it as a scale factor. Use a ruler that measures in centimeters.

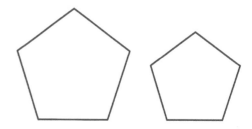

3. **a.** Are these two typing rates equal: 60 words per 90 seconds and 135 words per 3 minutes?

   b. If not, calculate how many more words you would type in 5 minutes at the faster rate than at the slower rate.

119

4. A family paid $40 for a meal in a restaurant. They gave the waiter a 5% tip (on the $40). They also paid a sales tax of 6.8% (on the $40) included in the total cost. What was the total cost they paid?

5. A ticket to ride a roller coaster costs $5 and a ticket to drive the bumper cars costs $4.50.

   a. How many percent more expensive is the first ticket than the second?

   b. How many percent cheaper is the second ticket than the first?

6. If the edges of this cube measure 2 inches, how many such cubes do you need in order to have 1 cubic foot?

7. Solve the equations.

a.	$\dfrac{x}{5} = -4.08$		**b.**	$\dfrac{w}{-0.2} = -0.4$
c.	$2x + 7 = -4(x + 5)$		**d.**	$\dfrac{x+1}{4} = -2$

8. The principal of a school wants to ask the students' parents whether some extra money should be used to purchase more books for the library, upgrade the computer systems, or to improve the sports facilities of the school. He randomly surveyed some parents who were attending a basketball game at the school.

Is the principal's sampling method biased or unbiased?

Explain why.

9. The dot plots show the amount of sugar in fruit and pop drinks (250 ml portion of drink).

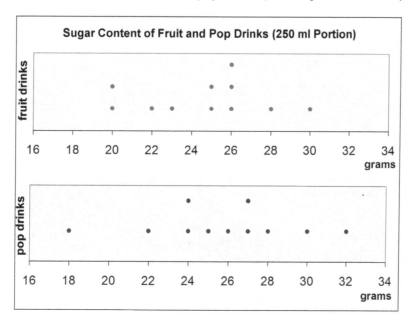

a. Based on the plots, which type of drink, if any, tends to contain more sugar?

Justify your answer.

b. Which type of drink, if any, has greater variability in the amount of sugar it has?

Justify your answer.

10. The map shows the center of Phoenix, Arizona. The distance AC along North 35th Avenue is 2 miles and the distance CB along West McDowell Road is also 2 miles.

    Calculate, to the hundredth of a mile, how much longer it is to travel from A to C and then from C to B than to travel from A to B along Grand Avenue.

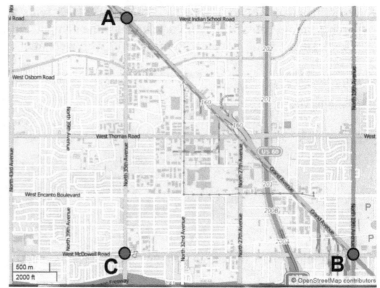

© OpenStreetMap contributors
Licensed under the license at
www.openstreetmap.org/copyright

11. A bag contains 5 blue socks, 2 red socks, and 6 white socks.

    a. You draw one randomly. What is the probability that the sock is white?

    b. The first sock was indeed white! You draw another one randomly. What is the probability that this one is white?

    c. You draw two socks randomly. What is the probability of getting two socks of the same color?